Student Edition

Eureka Math
Grade 2
Module 4

Special thanks go to the Gordon A. Cain Center and to the Department of Mathematics at Louisiana State University for their support in the development of *Eureka Math*.

For a free *Eureka Math* Teacher Resource Pack, Parent Tip Sheets, and more please visit www.Eureka.tools

ISBN 978-1-63255-294-5

Name _____   Date _____

1. Complete each *more* or *less* statement.

    a. 1 more than 66 is _____.

    b. 10 more than 66 is _____.

    c. 1 less than 66 is _____.

    d. 10 less than 66 is _____.

    e. 56 is 10 more than _____.

    f. 88 is 1 less than _____.

    g. _____ is 10 less than 67.

    h. _____ is 1 more than 72.

    i. 86 is _____ than 96.

    j. 78 is _____ than 79.

2. Circle the rule for each pattern.

    a. 34, 33, 32, 31, 30, 29   1 less   1 more   10 less   10 more

    b. 53, 63, 73, 83, 93   1 less   1 more   10 less   10 more

3. Complete each pattern.

    a. 37, 38, 39, _____, _____, _____

    b. 68, 58, 48, _____, _____, _____

    c. 51, 50, _____, _____, _____, 46

    d. 9, 19, _____, _____, _____, 59

Lesson 1:   Relate 1 more, 1 less, 10 more, and 10 less to addition and subtraction of 1 and 10.

©2015 Great Minds. eureka-math.org
G2-M4-SE-B2-1.3.1-1.2016

1

4. Complete each statement to show mental math using the arrow way.

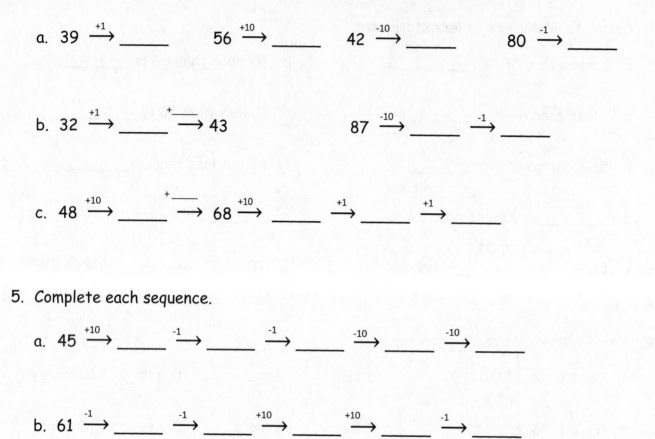

a. 39 $\xrightarrow{+1}$ _____   56 $\xrightarrow{+10}$ _____   42 $\xrightarrow{-10}$ _____   80 $\xrightarrow{-1}$ _____

b. 32 $\xrightarrow{+1}$ _____ $\xrightarrow{+\_\_}$ 43   87 $\xrightarrow{-10}$ _____ $\xrightarrow{-1}$ _____

c. 48 $\xrightarrow{+10}$ _____ $\xrightarrow{+\_\_}$ 68 $\xrightarrow{+10}$ _____ $\xrightarrow{+1}$ _____ $\xrightarrow{+1}$ _____

5. Complete each sequence.

a. 45 $\xrightarrow{+10}$ _____ $\xrightarrow{-1}$ _____ $\xrightarrow{-1}$ _____ $\xrightarrow{-10}$ _____ $\xrightarrow{-10}$ _____

b. 61 $\xrightarrow{-1}$ _____ $\xrightarrow{-1}$ _____ $\xrightarrow{+10}$ _____ $\xrightarrow{+10}$ _____ $\xrightarrow{-1}$ _____

6. Solve each word problem using the arrow way to record your mental math.

a. Yesterday Isaiah made 39 favor bags for his party. Today he made 23 more. How many favor bags did he make for his party?

b. There are 61 balloons. 12 blew away. How many are left?

Lesson 1:   Relate 1 more, 1 less, 10 more, and 10 less to addition and subtraction of 1 and 10.

©2015 Great Minds. eureka-math.org
G2-M4-SE-B2-1.3.1-1.2016

EUREKA MATH™

Name _____ Date _____

1. Complete each *more* or *less* statement.

    a. 1 more than 37 is _____.

    b. 10 more than 37 is _____.

    c. 1 less than 37 is _____.

    d. 10 less than 37 is _____.

    e. 58 is 10 more than _____.

    f. 29 is 1 less than _____.

    g. _____ is 10 less than 45.

    h. _____ is 1 more than 38.

    i. 49 is _____ than 50.

    j. 32 is _____ than 22.

2. Complete each pattern and write the rule.

    a. 44, 45, _____, _____, 48   Rule: _____

    b. 44, _____, 24, _____, 4   Rule: _____

    c. 44, _____, _____, 74, 84   Rule: _____

    d. _____, 43, 42, _____, 40   Rule: _____

    e. _____, _____, 44, 34, _____   Rule: _____

    f. 41, _____, _____, 38, 37   Rule: _____

EUREKA MATH™

Lesson 1:   Relate 1 more, 1 less, 10 more, and 10 less to addition and subtraction of 1 and 10.

3

©2015 Great Minds. eureka-math.org
G2-M4-SE-B2-1.3.1-1.2016

3. Label each statement as true or false.

   a. 1 more than 36 is the same as 1 less than 38. _____

   b. 10 less than 47 is the same as 1 more than 35. _____

   c. 10 less than 89 is the same as 1 less than 90. _____

   d. 10 more than 41 is the same as 1 less than 43. _____

4. Below is a chart of balloons at the county fair.

Color of Balloons	Number of Balloons
Red	59
Yellow	61
Green	65
Blue	
Pink	

   a. Use the following to complete the chart and answer the question.
      ▪ The fair has 1 more blue than red balloons.
      ▪ There are 10 fewer pink than yellow balloons.

   Are there more blue or pink balloons?

   b. If 1 red balloon pops and 10 red balloons fly away, how many red balloons are left? Use the arrow way to show your work.

Lesson 1:   Relate 1 more, 1 less, 10 more, and 10 less to addition and subtraction of 1 and 10.

©2015 Great Minds. eureka-math.org
G2-M4-SE-B2-1.3.1-1.2016

unlabeled tens place value chart

Lesson 1: Relate 1 more, 1 less, 10 more, and 10 less to addition and subtraction of 1 and 10.

5

This page intentionally left blank

Name _____   Date _____

1. Solve using place value strategies. Use your personal white board to show the arrow way or number bonds, or just use mental math, and record your answers.

   a. 5 tens + 3 tens = _530_ tens   2 tens + 7 tens = _270_ tens

               50 + 30 = _80_   20 + 70 = _90_

   b. 24 + 30 = _64_   50 + 24 = _74_   14 + 50 = _64_

   c. 20 + 37 = _57_   37 + 40 = _77_   60 + 27 = _87_

   d. 57 + _30_ = 87   _40_ + 34 = 74   19 + ____ = 69

   e. ____ + 56 = 86   38 + ____ = 78   12 + ____ = 72

2. Solve using place value strategies.

   a. 8 tens – 2 tens = _____ tens   7 tens – 3 tens = _____ tens

               80 – 20 = _____   70 – 30 = _____

   b. 78 - 40 = _____   56 – 30 = _____   88 – 50 = _____

   c. 84 - ____ = 24   57 - ____ = 37   93 - ____ = 43

   d. 83 - ____ = 23   54 - ____ = 34   91 - ____ = 41

3. Solve.

    a. 39 + _____ = 69

    b. 8 tens 7 ones – 3 tens = _____

    c. _____ + 5 tens = 7 tens

    d. _____ + 5 tens 6 ones = 8 tens 6 ones

    e. 48 ones – 2 tens = _____ tens _____ ones

4. Mark had 78 puzzle pieces. He lost 30 pieces. How many pieces does Mark have left? Use the arrow way to show your simplifying strategy.

Lesson 2:   Add and subtract multiples of 10 including counting on to subtract.

EUREKA MATH

Name _____ Date _____

1. Solve using place value strategies. Use scrap paper to show the arrow way or number bonds, or just use mental math, and record your answers.

a. 2 tens + 3 tens = __230__ tens
 	
     20 + 30 = __50__
 	
2 tens 4 ones + 3 tens = _9_ tens _2_ ones
 	
     24 + 30 = __54__
	b. 5 tens + 4 tens = __540__ tens
	
	     50 + 40 = __90__
	
	5 tens 9 ones + 4 tens = _9_ tens _9_ ones
	
	     59 + 40 = __99__

c. 28 + 40 = _____   18 + 30 = _____   60 + 38 = _____

d. 30 + 25 = _____   35 + 50 = _____   15 + 20 = _____

e. 37 + _____ = 47   _____ + 27 = 57   17 + _____ = 87

f. _____ + 22 = 62   29 + _____ = 79   11 + _____ = 91

2. Find each sum. Then use >, <, or = to compare.

a. 23 + 40 _____ 20 + 33

b. 50 + 18 _____ 48 + 20

c. 19 + 60 _____ 39 + 30

d. 64 + 10 _____ 49 + 20

e. 70 + 21 _____ 18 + 80

f. 35 + 50 _____ 26 + 60

Lesson 2:   Add and subtract multiples of 10 including counting on to subtract.

9

3. Solve using place value strategies.

a. 6 tens – 2 tens = ____ tens
 	
    60 – 20 = _____
 	
6 tens 3 ones – 3 tens = __ tens __ ones
 	
    63 – 30 = _____
	b. 8 tens – 5 tens = ____ tens
	
	    80 – 50 = _____
	
	8 tens 9 ones – 5 tens = __ tens __ ones
	
	    89 – 50 = _____

c. 55 – 20 = _____   75 – 30 = _____   85 – 50 = _____

d. 72 – ____ = 22   49 – ____ = 19   88 – ____ = 28

e. 67 – ____ = 47   71 – ____ = 51   99 – ____ = 69

4. Complete each more than or less than statement.

   a. 20 less than 58 is _____.   b. 36 more than 40 is _____.

   c. 40 less than _____ is 28.   d. 50 more than _____ is 64.

5. There were 68 plates in the sink at the end of the day. There were 40 plates in the sink at the beginning of the day. How many plates were added throughout the day? Use the arrow way to show your simplifying strategy.

Lesson 2:   Add and subtract multiples of 10 including counting on to subtract.

EUREKA
MATH™

Name _____   Date _____

1. Solve. Draw and label a tape diagram to subtract tens. Write the new number sentence.

   a. 23 – 9 = ___24 – 10___ = _____

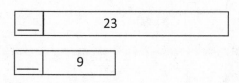

   b. 32 – 19 = _____ = _____

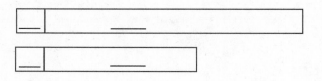

   c. 50 – 29 = _____ = _____

   d. 47 – 28 = _____ = _____

2. Solve. Draw and label a tape diagram to add tens. Write the new number sentence.

a. 29 + 46 = ___30 + 45___ = _____

29	1	45

b. 38 + 45 = _____ = _____

c. 61 + 29 = _____ = _____

d. 27 + 68 = _____ = _____

Lesson 4:   Add and subtract multiples of 10 and some ones within 100.

©2015 Great Minds. eureka-math.org
G2-M4-SE-B2-1.3.1-1.2016

Name _____ Date _____

1. Solve. Draw and label a tape diagram to subtract 10, 20, 30, 40, etc.

a. 17 – 9 = ___18 – 10___ = _____

```
                  18
        ┌──────────────────────┐
    ┌──┬─────────────────────┐
    │+1│          17          │
    └──┴─────────────────────┘
    ┌──┬──────────┐
    │+1│     9     │
    └──┴──────────┘
    └──────────┘  └──────────┘
         10            ?
```

b. 33 – 19 = _____ = _____

```
    ┌──┬──────────────────────────────────┐
    │__│          ___                       │
    └──┴──────────────────────────────────┘
    ┌──┬──────────────────┐
    │__│        ___         │
    └──┴──────────────────┘
```

c. 60 – 29 = _____ = _____

d. 56 – 38 = _____ = _____

2. Solve. Draw a number bond to add 10, 20, 30, 40, etc.

a. 28 + 43 = ___30 + 41___ = _____
                    /\
                  2 41

b. 49 + 26 = _____ = _____

c. 43 + 19 = _____ = _____

d. 67 + 28 = _____ = _____

3. Kylie has 28 more oranges than Cynthia. Kylie has 63 oranges. How many oranges does Cynthia have? Draw a tape diagram or number bond to solve.

Name _____   Date _____

Solve and show your strategy.

1. 39 books were on the top bookshelf. Marcy added 48 more books to the top shelf.
    How many books are on the top shelf now?

2. There are 53 regular pencils and some colored pencils in the bin. There are a total
    of 91 pencils in the bin. How many colored pencils are in the bin?

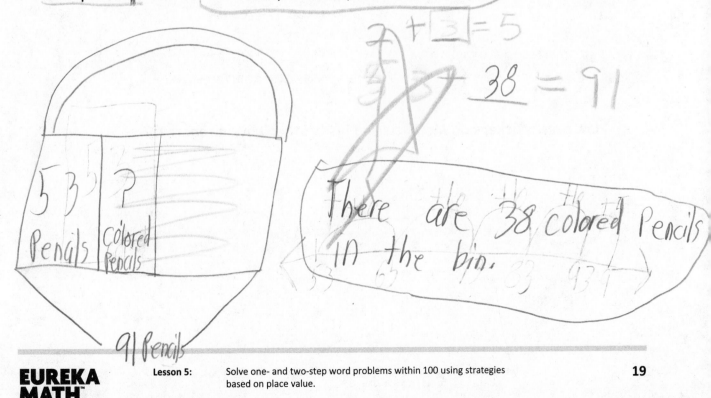

EUREKA MATH™

Lesson 5:   Solve one- and two-step word problems within 100 using strategies
             based on place value.

©2015 Great Minds. eureka-math.org
G2-M4-SE-B2-1.3.1-1.2016

19

3. Henry solved 24 of his homework problems. There were 51 left to do. How many math problems were there on his homework sheet?

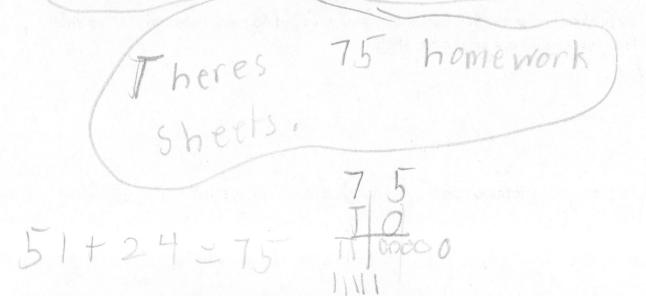

Theres 75 homework sheets.

51 + 24 = 75

4. Matthew has 68 stickers. His brother has 29 fewer stickers.

   a. How many stickers does Matthew's brother have?

   68 - 29 = 21

   Matthews has 21 stickers!

   b. How many stickers do Matthew and his brother have altogether?

5. There are 47 photos in the blue album. The blue album has 32 more photos than the red album.

   a. How many photos are in the red album?

   b. How many photos are in the red and blue albums altogether?

6. Kiera has 62 blocks, and Pete has 37 blocks. They give away 75 blocks. How many blocks do they have left?

Lesson 5:   Solve one- and two-step word problems within 100 using strategies based on place value.

©2015 Great Minds. eureka-math.org
G2-M4-SE-B2-1.3.1-1.2016

21

Name _____ Date _____

Solve and show your strategy.

1. 38 markers were in the bin. Chase added the 43 markers that were on the floor to the bin. How many markers are in the bin now?

2. There are 29 fewer big stickers on the sticker sheet than little stickers. There are 62 little stickers on the sheet. How many big stickers are there?

Lesson 5:   Solve one- and two-step word problems within 100 using strategies based on place value.

3. Rose has 34 photos in a photo album and 41 photos in a box. How many photos does Rose have?

4. Halle has two ribbons. The blue ribbon is 58 cm. The green ribbon is 38 cm longer than the blue ribbon.

    a. How long is the green ribbon?

    b. Halle uses 67 cm of green ribbon to wrap a present. How much green ribbon is left?

EUREKA
MATH™

Lesson 5:   Solve one- and two-step word problems within 100 using strategies
              based on place value.

©2015 Great Minds. eureka-math.org
G2-M4-SE-B2-1.3.1-1.2016

23

5. Chad bought a shirt for $19 and a pair of shoes for $28 more than the shirt.

   a. How much was the pair of shoes?

   b. How much money did Chad spend on the shirt and shoes?

   c. If Chad had $13 left over, how much money did Chad have before buying the shirt and shoes?

Name _____   Date _____

1. Solve using mental math, if you can. Use your place value chart and place value disks to solve those you cannot solve mentally.

    a. 6 + 8 = _____   30 + 8 = _____   36 + 8 = _____   36 + 48 = _____

    b. 5 + 7 = _____   20 + 7 = _____   25 + 7 = _____   25 + 57 = _____

2. Solve the following problems using your place value chart and place value disks. Compose a ten, if needed. Think about which ones you can solve mentally, too!

    a. 35 + 5 = _____   35 + 6 = _____

    b. 26 + 4 = _____   26 + 5 = _____

    c. 54 + 15 = _____   54 + 18 = _____

    d. 67 + 23 = _____   67 + 25 = _____

    e. 45 + 26 = _____   45 + 23 = _____

    f. 58 + 23 = _____   58 + 25 = _____

    g. 49 + 37 = _____   52 + 36 = _____

Lesson 6:   Use manipulatives to represent the composition of 10 ones as 1 ten with two-digit addends.

25

©2015 Great Minds. eureka-math.org
G2-M4-SE-B2-1.3.1-1.2016

3. There are 47 blue buttons and 25 black buttons in Sean's drawer. How many buttons are in his drawer?

For early finishers:

4. Leslie has 24 blue and 24 pink hair ribbons. She buys 17 more blue ribbons and 13 more pink ribbons from the store.

    a. How many blue hair ribbons does she have now?

    b. How many pink hair ribbons does she have now?

    c. Jada has 29 more pink ribbons than Leslie. How many pink ribbons does Jada have?

Lesson 6:   Use manipulatives to represent the composition of 10 ones as 1 ten with two-digit addends.

©2015 Great Minds. eureka-math.org
G2-M4-SE-B2-1.3.1-1.2016

Name _____   Date _____

1. Solve using mental math, if you can. Use your place value chart and place value disks to solve those you cannot do mentally.

   a. 4 + 9 = _____   30 + 9 = _____   34 + 9 = _____   34 + 49 = _____

   b. 6 + 8 = _____   20 + 8 = _____   26 + 8 = _____   26 + 58 = _____

2. Solve the following problems using your place value chart and place value disks. Compose a ten, if needed. Think about which ones you can solve mentally, too!

   a. 21 + 9 = _____   22 + 9 = _____

   b. 28 + 2 = _____   28 + 4 = _____

   c. 32 + 16 = _____   34 + 17 = _____

   d. 47 + 23 = _____   47 + 25 = _____

   e. 53 + 35 = _____   58 + 35 = _____

   f. 58 + 42 = _____   58 + 45 = _____

   g. 69 + 32 = _____   36 + 62 = _____

   h. 77 + 13 = _____   16 + 77 = _____

   i. 59 + 34 = _____   31 + 58 = _____

Lesson 6:   Use manipulatives to represent the composition of 10 ones as 1 ten   27
             with two-digit addends.

©2015 Great Minds. eureka-math.org
G2-M4-SE-B2-1.3.1-1.2016

Solve using a place value chart.

3. Melissa has 36 more crayons than her brother. Her brother has 49 crayons.
    How many crayons does Melissa have?

4. There were 67 candles on Grandma's birthday cake and 26 left in the box.
    How many candles were there in all?

5. Frank's mother gave him $25 to save. If he already had $38 saved, how much
    money does Frank have saved now?

Lesson 6:   Use manipulatives to represent the composition of 10 ones as 1 ten
with two-digit addends.

©2015 Great Minds. eureka-math.org
G2-M4-SE-B2-1.3.1-1.2016

Name _____ Date _____

1. Solve the following problems using the vertical form, your place value chart, and place value disks. Bundle a ten, when necessary. Think about which ones you can solve mentally, too!

   a. 22 + 8   21 + 9

   b. 34 + 17   33 + 18

   c. 48 + 34   46 + 36

   d. 27 + 68   26 + 69

Lesson 7:   Relate addition using manipulatives to a written vertical method.

29

©2015 Great Minds. eureka-math.org
G2-M4-SE-B2-1.3.1-1.2016

©2015 Great Minds. eureka-math.org
G2-M4-SE-B2-1.3.1-1.2016

Extra Practice for Early Finishers: Solve the following problems using your place value chart and place value disks. Bundle a ten, when necessary.

2. Samantha brought grapes to school for a snack. She had 27 green grapes and 58 red grapes. How many grapes did she bring to school?

3. Thomas read 29 pages of his new book on Monday. On Tuesday, he read 35 more pages than he did on Monday.

   a. How many pages did Thomas read on Tuesday?

   b. How many pages did Thomas read on both days?

Lesson 7: Relate addition using manipulatives to a written vertical method.

Name _____   Date _____

1. Solve the following problems using the vertical form, your place value chart, and place value disks. Bundle a ten, if needed. Think about which ones you can solve mentally, too!

   a. 31 + 9   32 + 8

   b. 42 + 18   43 + 17

   c. 26 + 67   28 + 65

2. Add the bottom numbers to find the missing number above it.

3. Jahsir counted 63 flowers by the door and 28 flowers on the windowsill. How many flowers were by the door and on the windowsill?

4. Antonio's string is 38 centimeters longer than his reading book. The length of his reading book is 26 centimeters.

   a. What is the length of Antonio's string?

   b. The length of Antonio's reading book is 20 centimeters shorter than the length of his desk. How long is Antonio's desk?

Name _____   Date _____

1. Solve vertically. Draw and bundle place value disks on the place value chart.

a. 27 + 15 = _____

b. 44 + 26 = _____

c. 48 + 31 = _____

d. 33 + 59 = _____

Lesson 8:   Use math drawings to represent the composition and relate drawings
to a written method.

©2015 Great Minds. eureka-math.org
G2-M4-SE-B2-1.3.1-1.2016

33

e. 27 + 45 = _____

f. 18 + 68 = _____

2. There are 23 laptops in the computer room and 27 laptops in the first-grade classroom. How many laptops are in the computer room and first-grade classroom altogether?

For early finishers:

3. Mrs. Anderson gave 36 pencils to her class and had 48 left over. How many pencils did Mrs. Anderson have at first?

Lesson 8:   Use math drawings to represent the composition and relate drawings to a written method.

Name _____   Date _____

1. Solve vertically. Draw and bundle place value disks on the place value chart.

    a. 26 + 35 = _____

    b. 28 + 14 = _____

    c. 35 + 27 = _____

    d. 23 + 46 = _____

EUREKA MATH

Lesson 8:   Use math drawings to represent the composition and relate drawings
                to a written method.

©2015 Great Minds. eureka-math.org
G2-M4-SE-B2-1.3.1-1.2016

35

e. $32 + 59 =$ _____

2. Twenty-eight second-grade students went on a field trip to the zoo. The other 24 second-grade students stayed at school. How many second-grade students are there in all?

3. Alice cut a 27-cm piece of ribbon and had 39 cm of ribbon left over. How much ribbon did Alice have at first?

    Lesson 8:   Use math drawings to represent the composition and relate drawings to a written method.

©2015 Great Minds. eureka-math.org
G2-M4-SE-B2-1.3.1-1.2016

Name _____   Date _____

1. Solve using the algorithm. Draw and bundle chips on the place value chart.

    a. 123 + 16 = _____

hundreds	tens	ones

    b. 111 + 79 = _____

hundreds	tens	ones

    c. 109 + 33 = _____

hundreds	tens	ones

EUREKA MATH™

Lesson 9: Use math drawings to represent the composition when adding a two-digit to a three digit addend.

37

d.   57 + 138 = _____

hundreds	tens	ones

2.   Jose sold 127 books in the morning. He sold another 35 books in the afternoon. At the end of the day he had 19 books left.

    a.   How many books did Jose sell?

hundreds	tens	ones

    b.   How many books did Jose have at the beginning of the day?

hundreds	tens	ones

Lesson 9:   Use math drawings to represent the composition when adding a two-digit to a three digit addend.

©2015 Great Minds. eureka-math.org
G2-M4-SE-B2-1.3.1-1.2016

EUREKA MATH™

Name _____   Date _____

1. Solve using the algorithm. Draw and bundle chips on the place value chart.

a. 127 + 14 = _____

hundreds	tens	ones

b. 135 + 46 = _____

hundreds	tens	ones

c. 108 + 37 = _____

hundreds	tens	ones

©2015 Great Minds. eureka-math.org
G2-M4-SE-B2-1.3.1-1.2016

2. Solve using the algorithm. Write a number sentence for the problem modeled on the place value chart.

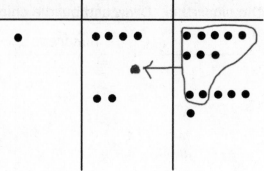

3. Jane made 48 lemon bars and 23 cookies.

a. How many lemon bars and cookies did Jane make?

hundreds	tens	ones

b. Jane made 19 more lemon bars. How many lemon bars does she have?

hundreds	tens	ones

Lesson 9:   Use math drawings to represent the composition when adding a two-digit to a three digit addend.

Name _____   Date _____

1. Solve using the algorithm. Draw chips and bundle when you can.

   a. 127 + 18 = _____

hundreds	tens	ones

   b. 136 + 16 = _____

hundreds	tens	ones

   c. 109 + 41 = _____

hundreds	tens	ones

   d. 29 + 148 = _____

hundreds	tens	ones

EUREKA MATH

Lesson 10:   Use math drawings to represent the composition when adding a
two-digit to a three digit addend.

41

©2015 Great Minds. eureka-math.org
G2-M4-SE-B2-1.3.1-1.2016

e. 79 + 107 = _____

hundreds	tens	ones

Before bundling a ten   _____ hundreds   _____ tens   _____ ones

After bundling a ten   _____ hundreds   _____ tens   _____ ones

2. a. On Saturday, Colleen earned 4 ten-dollar bills and 18 one-dollar bills working on the farm. How much money did Colleen earn?

hundreds	tens	ones

b. On Sunday, Colleen earned 2 ten-dollar bills and 16 one-dollar bills. How much money did she earn on both days?

hundreds	tens	ones

Lesson 10:   Use math drawings to represent the composition when adding a two-digit to a three digit addend.

EUREKA MATH

Name _____ Date _____

1. Solve using the algorithm. Draw chips and bundle when you can.

   a. 125 + 17 = _____

hundreds	tens	ones

   b. 148 + 14 = _____

hundreds	tens	ones

   c. 107 + 56 = _____

hundreds	tens	ones

   d. 38 + 149 = _____

hundreds	tens	ones

EUREKA MATH

Lesson 10: Use math drawings to represent the composition when adding a two-digit to a three digit addend.

43

©2015 Great Minds. eureka-math.org
G2-M4-SE-B2-1.3.1-1.2016

2. Jamie started to solve this problem when she accidentally dropped paint on her sheet. Can you figure out what problem she was given and her answer by looking at her work?

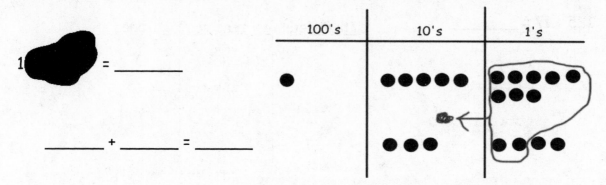

1 ⬛ = _____

_____ + _____ = _____

3. a. In the morning, Mateo borrowed 4 bundles of ten markers and 17 loose markers from the art teacher. How many markers did Mateo borrow?

hundreds	tens	ones

b. In the afternoon, Mateo borrowed 2 bundles of ten crayons and 15 loose crayons. How many markers and crayons did Mateo borrow in all?

hundreds	tens	ones

Lesson 10: Use math drawings to represent the composition when adding a two-digit to a three digit addend.

Name _____   Date _____

1. Solve using mental math.

    a. 8 – 7 = _____   38 – 7 = _____   38 – 8 = _____   38 – 9 = _____

    b. 7 – 6 = _____   87 – 6 = _____   87 – 7 = _____   87 – 8 = _____

2. Solve using your place value chart and place value disks. Unbundle a ten if needed.
    Think about which problems you can solve mentally, too!

    a. 28 – 7 = _____   28 – 9 = _____

    b. 25 – 5 = _____   25 – 6 = _____

    c. 30 – 5 = _____   33 – 5 = _____

    d. 47 – 22 = _____   41 – 22 = _____

    e. 44 – 16 = _____   44 – 26 = _____

    f. 70 – 28 = _____   80 – 28 = _____

Lesson 11:   Represent subtraction with and without the decomposition of   45
1 ten as 10 ones with manipulatives.

©2015 Great Minds. eureka-math.org
G2-M4-SE-B2-1.3.1-1.2016

3. Solve 56 – 28, and explain your strategy.

For early finishers:

4. There are 63 problems on the math test. Tamara answered 48 problems correctly, but the rest were incorrect. How many problems did she answer incorrectly?

5. Mr. Ross has 7 fewer students than Mrs. Jordan. Mr. Ross has 35 students. How many students does Mrs. Jordan have?

Lesson 11:   Represent subtraction with and without the decomposition of 1 ten as 10 ones with manipulatives.

Name _____ Date _____

1. Solve using mental math.

   a. 6 – 5 = _____   26 – 5 = _____   26 – 6 = _____   26 – 7 = _____

   b. 8 – 7 = _____   58 – 7 = _____   58 – 8 = _____   58 – 9 = _____

2. Solve using your place value chart and place value disks. Unbundle a ten, if needed. Think about which problems you can solve mentally, too!

   a. 36 – 5 = _____   36 – 7 = _____

   b. 37 – 6 = _____   37 – 8 = _____

   c. 40 – 5 = _____   41 – 5 = _____

   d. 58 – 32 = _____   58 – 29 = _____

   e. 60 – 26 = _____   62 – 26 = _____

   f. 70 – 41 = _____   80 – 41 = _____

Lesson 11:   Represent subtraction with and without the decomposition of 1 ten as 10 ones with manipulatives.

©2015 Great Minds. eureka-math.org
G2-M4-SE-B2-1.3.1-1.2016

47

3. Solve and explain your strategy.

a.

   41 – 27 = _____

b.

   67 – 28 = _____

4. The number of marbles in each jar is marked on the front. Miss Clark took
    37 marbles out of each jar. How many marbles are left in each jar? Complete the
    number sentence to find out.

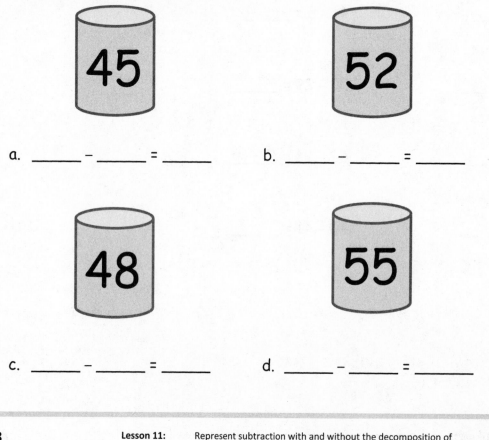

a. _____ – _____ = _____   b. _____ – _____ = _____

c. _____ – _____ = _____   d. _____ – _____ = _____

Lesson 11:   Represent subtraction with and without the decomposition of
                            1 ten as 10 ones with manipulatives.

©2015 Great Minds. eureka-math.org
G2-M4-SE-B2-1.3.1-1.2016

Name _____   Date _____

1. Use place value disks to solve each problem. Rewrite the problem vertically, and
    record each step as shown in the example.

    a. 22 – 18

$$
\begin{array}{r}
{\scriptstyle 1\ 12} \\
\cancel{22} \\
-\ 18 \\
\hline
4
\end{array}
$$

    b. 20 – 12

    c. 34 – 25   d. 25 – 18

    e. 53 – 29   f. 71 – 27

2. Terry and Pam both solved the problem 64 – 49. They came up with different answers and cannot agree on who is correct. Terry answered 25, and Pam answered 15. Use place value disks to explain who is correct, and rewrite the problem vertically to solve.

For early finishers:

3. Samantha has 42 marbles, and Graham has 17 marbles.

   a. How many more marbles does Samantha have than Graham?

   b. James has 25 fewer marbles than Samantha. How many marbles does James have?

Lesson 12:   Relate manipulative representations to a written method.

Name _____   Date _____

1. Use place value disks to solve each problem. Rewrite the problem vertically, and record each step as shown in the example.

    a. 34 – 18

$$
\begin{array}{r}
{}^{2}\;{}^{14}\!\!\!\!\!\! \\
3\,4 \\
-\ 1\,8 \\
\hline
1\,6
\end{array}
$$

    b. 41 – 16

    c. 33 – 15

    d. 46 – 18

    e. 62 – 27

    f. 81 – 34

2. Some first- and second-grade students voted on their favorite drink. The table shows the number of votes for each drink.

Types of Drink	Number of Votes
Milk	28
Apple Juice	19
Grape Juice	16
Fruit Punch	37
Orange Juice	44

a. How many more students voted for fruit punch than for milk? Show your work.

b. How many more students voted for orange juice than for grape juice? Show your work.

c. How many fewer students voted for apple juice than for milk? Show your work.

©2015 Great Minds. eureka-math.org
G2-M4-SE-B2-1.3.1-1.2016

Name _____ Date _____

1. Solve vertically. Draw a place value chart and chips to model each problem.
    Show how you change 1 ten for 10 ones, when necessary.

a. 31 – 19 = _____	b. 46 – 24 = _____
c. 51 – 33 = _____	d. 67 – 49 = _____
e. 66 – 48 = _____	f. 77 – 58 = _____

Lesson 13: Use math drawings to represent subtraction with and without
decomposition and relate drawings to a written method.

©2015 Great Minds. eureka-math.org
G2-M4-SE-B2-1.3.1-1.2016

53

2. Solve 31 – 27 and 25 – 15 vertically using the space below. Circle to tell if the number sentence is true or false.

True or False

31 – 27 = 25 – 15

3. Solve 78 – 43 and 81 – 46 vertically using the space below. Circle to tell if the number sentence is true or false.

True or False

78 – 43 = 81 – 46

4. Mrs. Smith has 39 tomatoes in her garden. Mrs. Thompson has 52 tomatoes in her garden. How many fewer tomatoes does Mrs. Smith have than Mrs. Thompson?

Lesson 13:   Use math drawings to represent subtraction with and without decomposition and relate drawings to a written method.

©2015 Great Minds. eureka-math.org
G2-M4-SE-B2-1.3.1-1.2016

Name _____   Date _____

1. Solve vertically. Use the place value chart and chips to model each problem.
    Show how you change 1 ten for 10 ones, when necessary. The first one has been
    started for you.

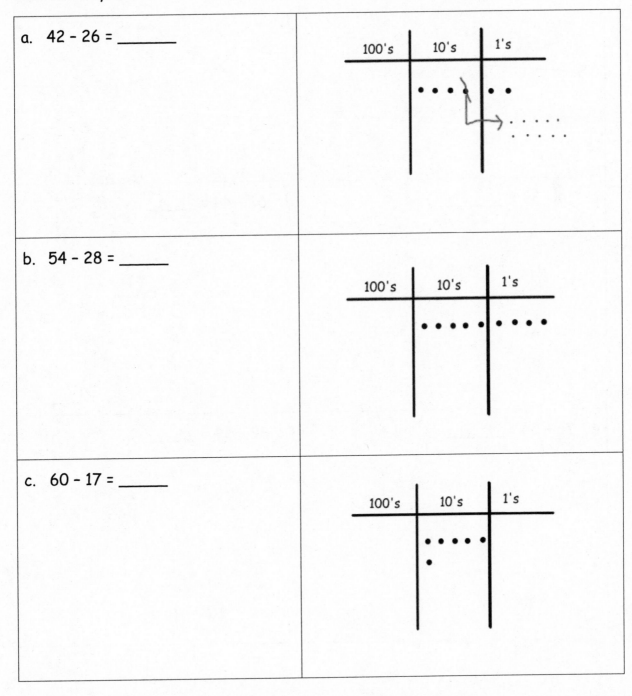

a. 42 – 26 = _____

b. 54 – 28 = _____

c. 60 – 17 = _____

Lesson 13:   Use math drawings to represent subtraction with and without
decomposition and relate drawings to a written method.

55

©2015 Great Minds. eureka-math.org
G2-M4-SE-B2-1.3.1-1.2016

2. Solve vertically. Draw a place value chart and chips to model each problem. Show how you change 1 ten for 10 ones, when necessary.

a. 31 – 19 = _____

b. 47 – 24 = _____

c. 51 – 39 = _____

d. 67 – 44 = _____

e. 76 – 54 = _____

f. 82 – 59 = _____

Lesson 13: Use math drawings to represent subtraction with and without decomposition and relate drawings to a written method.

©2015 Great Minds. eureka-math.org
G2-M4-SE-B2-1.3.1-1.2016

EUREKA MATH™

Name _____   Date _____

1. Solve by writing the problem vertically. Check your result by drawing chips on the place value chart. Change 1 ten for 10 ones, when needed.

a. 134 – 23= _____

hundreds	tens	ones

b. 140 – 12 = _____

hundreds	tens	ones

c. 121 – 14 = _____

hundreds	tens	ones

d. 161 – 26 = _____

hundreds	tens	ones

e. 187 – 49 = _____

hundreds	tens	ones

2. Solve the following problems vertically without a place value chart.

a. 63 – 28 = _____

b. 163 – 28 = _____

Lesson 14: Represent subtraction with and without the decomposition when there is a three-digit minuend.

EUREKA MATH

Name _____   Date _____

1. Solve by writing the problem vertically. Check your result by drawing chips on the place value chart. Change 1 ten for 10 ones, when needed.

a. 156 – 42 = _____

hundreds	tens	ones

b. 150 – 36 = _____

hundreds	tens	ones

c. 163 – 45 = _____

hundreds	tens	ones

Lesson 14:   Represent subtraction with and without the decomposition when there is a three-digit minuend.

59

2. Solve the following problems without a place value chart.

a.	b.
1 3 4 - 2 9	1 5 4 - 3 7

3. Solve and show your work. Draw a place value chart and chips, if needed.

   a. Aniyah has 165 seashells. She has 28 more than Ralph. How many seashells does Ralph have?

   b. Aniyah and Ralph each give 19 seashells to Harold. How many seashells does Aniyah have left?

   c. How many seashells does Ralph have left?

Lesson 14: Represent subtraction with and without the decomposition when there is a three-digit minuend.

©2015 Great Minds. eureka-math.org
G2-M4-SE-B2-1.3.1-1.2016

EUREKA MATH

Name _____   Date _____

1. Solve each problem using vertical form. Show the subtraction on the place value chart with chips. Exchange 1 ten for 10 ones, when necessary.

   a. 173 – 42

hundreds	tens	ones

   b. 173 – 38

hundreds	tens	ones

   c. 170 – 44

hundreds	tens	ones

EUREKA MATH™

Lesson 15:   Represent subtraction with and without the decomposition when there is a three-digit minuend.

©2015 Great Minds. eureka-math.org
G2-M4-SE-B2-1.3.1-1.2016

61

d. 150 – 19

hundreds	tens	ones

e. 186 – 57

hundreds	tens	ones

2. Solve the following problems without using a place value chart.

a. 73 – 56	b. 170 – 53

Name _____   Date _____

1. Solve each problem using vertical form. Show the subtraction on the place value chart with chips. Exchange 1 ten for 10 ones, when necessary.

    a. 153 – 31

hundreds	tens	ones

    b. 153 – 38

hundreds	tens	ones

    c. 160 – 37

hundreds	tens	ones

Lesson 15:   Represent subtraction with and without the decomposition when
there is a three-digit minuend.

©2015 Great Minds. eureka-math.org
G2-M4-SE-B2-1.3.1-1.2016

63

d. 182 – 59

hundreds	tens	ones

2. Lisa solved 166 – 48 vertically and on her place value chart. Explain what Lisa did correctly and what she needs to fix.

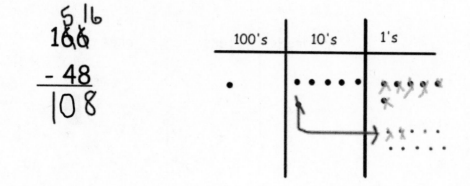

a. Lisa correctly _____

_____

b. Lisa needs to fix _____

_____

Lesson 15:   Represent subtraction with and without the decomposition when
                              there is a three-digit minuend.

EUREKA MATH™

Name _____   Date _____

Solve the following word problems. Use the RDW process.

1. Frederick counted a total of 80 flowers in the garden. There were 39 white flowers, and the rest were pink. How many flowers were pink?

2. The clothing store had 42 shirts. After selling some, there were 16 left. How many shirts were sold?

3. There were 26 magazines on Shelf A and 60 magazines on Shelf B. How many more magazines were on Shelf B than Shelf A?

Lesson 16:   Solve one- and two-step word problems within 100 using strategies based on place value.

65

©2015 Great Minds. eureka-math.org
G2-M4-SE-B2-1.3.1-1.2016

4. Andy spent 71 hours studying in November.

   In December, he studied 19 hours less.
   Rachel studied 22 hours more than Andy studied in December.
   How many hours did Rachel study in December?

5. Thirty-six books are in the blue bin.

   The blue bin has 18 more books than the red bin.
   The yellow bin has 7 more books than the red bin.

   a. How many books are in the red bin?

   b. How many books are in the yellow bin?

Lesson 16:   Solve one- and two-step word problems within 100 using strategies
                based on place value.

©2015 Great Minds. eureka-math.org
G2-M4-SE-B2-1.3.1-1.2016

Name _____   Date _____

Solve the following word problems. Use the RDW process.

1. Vicki modeled the following problem with a tape diagram.

    Eighty-two students are in the math club. 35 students are in the science club. How many more students are in the math club than science club?

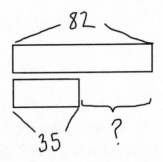

    Show another model to solve the problem. Write your answer in a sentence.

Lesson 16: Solve one- and two-step word problems within 100 using strategies based on place value.

67

©2015 Great Minds. eureka-math.org
G2-M4-SE-B2-1.3.1-1.2016

2. Forty-six birds sat on a wire. Some flew away, but 29 stayed. How many birds flew away? Show your work.

3. Ian bought a pack of 47 water balloons. 19 were red, 16 were yellow, and the rest were blue. How many water balloons were blue? Show your work.

4. Daniel read 54 pages of his book in the morning. He read 27 fewer pages in the afternoon. How many pages did Daniel read altogether? Show your work.

Lesson 16:   Solve one- and two-step word problems within 100 using strategies
                  based on place value.

©2015 Great Minds. eureka-math.org
G2-M4-SE-B2-1.3.1-1.2016

Name _____   Date _____

1. Solve mentally.

    a. 2 ones + _____ = 1 ten   2 + _____ = 10

        2 tens + _____ = 1 hundred   20 + _____ = 100

    b. 1 ten = _____ + 6 ones   10 = _____ + 6

        1 hundred = _____ + 6 tens   100 = _____ + 60

    c. 3 ones + 7 ones = _____ ten   3 + 7 = _____

        3 tens + 7 tens = _____ tens   30 + 70 = _____

        13 tens + 7 tens = _____ tens   130 + 70 = _____

    d. 6 ones + 4 ones = _____ ten   6 + 4 = _____

        16 tens + 4 tens = _____ hundreds   160 + 40 = _____

    e. 12 ones + 8 ones = _____ tens   12 + 8 = _____

        12 tens + 8 tens = _____ hundreds   120 + 80 = _____

EUREKA MATH™

Lesson 17:   Use mental strategies to relate compositions of 10 tens as
1 hundred to 10 ones as 1 ten.

69

©2015 Great Minds. eureka-math.org
G2-M4-SE-B2-1.3.1-1.2016

2. Solve.

a. 9 ones + 4 ones = _____ ten _____ ones   9 + 4 = _____

   9 tens + 4 tens = _____ hundred _____ tens   90 + 40 = _____

b. 4 ones + 8 ones = _____ ten _____ ones   4 + 8 = _____

   4 tens + 8 tens = _____ hundred _____ tens   40 + 80 = _____

c. 6 ones + 7 ones = _____ ten _____ ones   6 + 7 = _____

   6 tens + 7 tens = _____ hundred _____ tens   60 + 70 = _____

3. Fill in the blanks. Then, complete the addition sentence.
   The first one is done for you.

   a. $24 \xrightarrow{+6}$ _30_ $\xrightarrow{+70}$ _100_   b. $124 \xrightarrow{+6}$ _____ $\xrightarrow{+70}$ _____

      24 + _76_ = _100_   124 + _____ = _____

   c. $7 \xrightarrow{+3}$ _____ $\xrightarrow{+90}$ _____ $\xrightarrow{+100}$ _____   d. $70 \xrightarrow{+30}$ _____ $\xrightarrow{+90}$ _____ $\xrightarrow{+10}$ _____

      7 + _____ = _____   70 + _____ = _____

   e. $38 \xrightarrow{+2}$ _____ $\xrightarrow{+60}$ _____ $\xrightarrow{+30}$ _____   f. $98 \xrightarrow{+2}$ _____ $\xrightarrow{+6}$ _____ $\xrightarrow{+40}$ _____

      38 + _____ = _____   98 + _____ = _____

Lesson 17:   Use mental strategies to relate compositions of 10 tens as
               1 hundred to 10 ones as 1 ten.

Name _____ Date _____

1. Solve mentally.

   a. 4 ones + _____ = 1 ten   4 + _____ = 10

      4 tens + _____ = 1 hundred   40 + _____ = 100

   b. 1 ten = _____ + 7 ones   10 = _____ + 7

      1 hundred = _____ + 7 tens   100 = _____ + 70

   c. 1 ten more than 9 ones = _____   10 + 9 = _____

      1 hundred more than 9 ones = _____   100 + 9 = _____

      1 hundred more than 9 tens = _____   100 + 90 = _____

   d. 2 ones + 8 ones = _____ ten   2 + 8 = _____

      2 tens + 8 tens = _____ hundred   20 + 80 = _____

   e. 5 ones + 6 ones = ____ten(s) ____ one(s)   5 + 6 = _____

      5 tens + 6 tens = ____hundred(s) ____ ten(s)   50 + 60 = _____

   f. 14 ones + 4 ones = ____ ten(s) ____ one(s)   14 + 4 = _____

      14 tens + 4 tens = ____ hundred(s) ____ tens(s)   140 + 40 = _____

Lesson 17:   Use mental strategies to relate compositions of 10 tens as 1 hundred to 10 ones as 1 ten.

71

2. Solve.

    a.   6 ones + 5 ones = _____ ten _____ one   6 + 5 = _____

        6 tens + 5 tens = _____ hundred _____ ten   60 + 50 = _____

    b.   5 ones + 7 ones = _____ ten _____ ones   5 + 7 = _____

        5 tens + 7 tens = _____ hundred _____ tens   50 + 70 = _____

    c.   9 ones + 8 ones = _____ ten _____ ones   9 + 8 = _____

        9 tens + 8 tens = _____ hundred _____ tens   90 + 80 = _____

3. Fill in the blanks. Then, complete the addition sentence. The first one is done for you.

    a. $36 \xrightarrow{+4}$ __40__ $\xrightarrow{+60}$ __100__ $\xrightarrow{+30}$ __130__   b. $78 \xrightarrow{+2}$ _____ $\xrightarrow{+10}$ _____ $\xrightarrow{+10}$ _____

        36 + __94__ = __130__   78 + _____ = _____

    c. $61 \xrightarrow{+9}$ _____ $\xrightarrow{+10}$ _____ $\xrightarrow{+10}$ _____ $\xrightarrow{+10}$ _____ $\xrightarrow{+100}$ _____

        61 + _____ = _____

    d. $27 \xrightarrow{+3}$ _____ $\xrightarrow{+70}$ _____ $\xrightarrow{+100}$ _____

        27 + _____ = _____

       Lesson 17:   Use mental strategies to relate compositions of 10 tens as
                             1 hundred to 10 ones as 1 ten.

©2015 Great Minds. eureka-math.org
G2-M4-SE-B2-1.3.1-1.2016

Name _____   Date _____

1. Solve using your place value chart and place value disks.

    a.   80 + 30 = _____   90 + 40 = _____

    b.   73 + 38 = _____   73 + 49 = _____

    c.   93 + 38 = _____   42 + 99 = _____

    d.   84 + 37 = _____   69 + 63 = _____

    e.   113 + 78 = _____   128 + 72 = _____

2. Circle the statements that are true as you solve each problem using place value disks.

a. 47 + 123	b. 97 + 54
I change 10 ones for 1 ten.	I change 10 ones for 1 ten.
I change 10 tens for 1 hundred.	I change 10 tens for 1 hundred.
The total of the two parts is 160.	The total of the two parts is 141.
The total of the two parts is 170.	The total of the two parts is 151.

Lesson 18:   Use manipulatives to represent additions with two compositions.

73

©2015 Great Minds. eureka-math.org
G2-M4-SE-B2-1.3.1-1.2016

3. Write an addition sentence that corresponds to the following number bond. Solve the problem using your place value disks, and fill in the missing total.

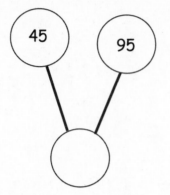

4. There are 50 girls and 80 boys in the after school program. How many children are in the after school program?

5. Kim and Stacy solved 83 + 39. Kim's answer was less than 120. Stacy's answer was more than 120. Whose answer was incorrect? Explain how you know using words, pictures, or numbers.

Name _____   Date _____

1. Solve using your place value chart and place value disks.

    a.   20 + 90 = _____   60 + 70 = _____

    b.   29 + 93 = _____   69 + 72 = _____

    c.   45 + 86 = _____   46 + 96 = _____

    d.   47 + 115 = _____   47 + 95 = _____

    e.   28 + 72 = _____   128 + 72 = _____

2. Circle the statements that are true as you solve each problem using place value disks.

a. 68 + 51	b. 127 + 46
I change 10 ones for 1 ten.	I change 10 ones for 1 ten.
I change 10 tens for 1 hundred.	I change 10 tens for 1 hundred.
The total of the two parts is 109.	The total of the two parts is 163.
The total of the two parts is 119.	The total of the two parts is 173.

3. Solve the problem using your place value disks, and fill in the missing total. Then, write an addition sentence that relates to the number bonds.

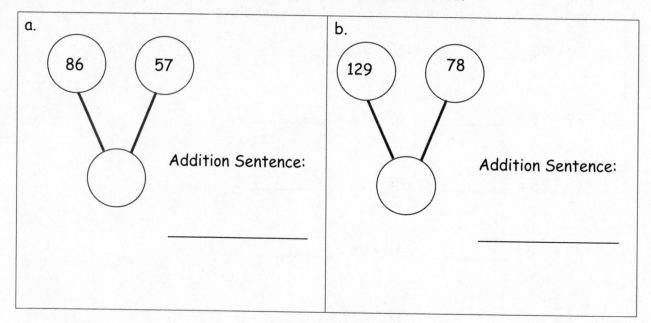

a.

86   57

Addition Sentence:

_____

b.

129   78

Addition Sentence:

_____

4. Solve using your place value chart and place value disks.

a.   45 + 55 = _____

b.   78 + 33 = _____

c.   37 + 84 = _____

Lesson 18:   Use manipulatives to represent additions with two compositions.

unlabeled hundreds place value chart

Lesson 18: Use manipulatives to represent additions with two compositions.

77

©2015 Great Minds. eureka-math.org
G2-M4-SE-B2-1.3.1-1.2016

This page intentionally left blank

Name _____   Date _____

1. Solve the following problems using the vertical form, your place value chart, and place value disks. Bundle a ten or hundred, if needed.

a. 72 + 19	b. 28 + 91
c. 68 + 61	d. 97 + 35
e. 68 + 75	f. 96 + 47

©2015 Great Minds. eureka-math.org
G2-M4-SE-B2-1.3.1-1.2016

g. 177 + 23	h. 146 + 54

2. Thirty-eight fewer girls attended summer camp than boys. Seventy-nine girls attended.

a.   How many boys attended summer camp?

b.   How many children attended summer camp?

Name _____   Date _____

1. Solve the following problems using the vertical form, your place value chart, and place value disks. Bundle a ten or hundred, if needed.

a. 84 + 37	b. 42 + 79
c. 58 + 56	d. 46 + 96
e. 75 + 69	f. 48 + 94

©2015 Great Minds. eureka-math.org
G2-M4-SE-B2-1.3.1-1.2016

g. 162 + 38	h. 156 + 44

2. Seventy-four trees were planted in the garden. Forty-nine more bushes were planted than trees in the garden.

   a.   How many bushes were planted?

   b.   How many trees and bushes were planted?

Lesson 19:   Relate manipulative representations to a written method.

©2015 Great Minds. eureka-math.org
G2-M4-SE-B2-1.3.1-1.2016

Name _____   Date _____

1. Solve vertically. Draw chips on the place value chart and bundle, when needed.

    a.   23 + 57 = _____

100's	10's	1's

    b.   65 + 36 = _____

100's	10's	1's

    c.   83 + 29 = _____

100's	10's	1's

Lesson 20:   Use math drawings to represent additions with up to two
compositions and relate drawings to a written method.

83

©2015 Great Minds. eureka-math.org
G2-M4-SE-B2-1.3.1-1.2016

d. 47 + 75 = _____

100's	10's	1's

e. 68 + 88 = _____

100's	10's	1's

2. Jessica's teacher marked her work incorrect for the following problem. Jessica cannot figure out what she did wrong. If you were Jessica's teacher, how would you explain her mistake?

Jessica's work:	Explanation:
100's 10's 1's 77 +32 19	

Lesson 20:   Use math drawings to represent additions with up to two
compositions and relate drawings to a written method.

EUREKA MATH

Name _____ Date _____

1. Solve vertically. Draw chips on the place value chart and bundle, when needed.

   a. 41 + 39 = _____

100's	10's	1's

   b. 54 + 26 = _____

100's	10's	1's

   c. 96 + 39 = _____

100's	10's	1's

EUREKA MATH™

Lesson 20: Use math drawings to represent additions with up to two compositions and relate drawings to a written method.

©2015 Great Minds. eureka-math.org
G2-M4-SE-B2-1.3.1-1.2016

85

d. 84 + 79 = _____

100's	10's	1's

e. 65 + 97 = _____

100's	10's	1's

2. For each box, find and circle two numbers that add up to 150.

a.		b.		c.	
67	63	48	92	75	55
73	83	68	62	65	45
57		58		75	

Lesson 20:   Use math drawings to represent additions with up to two
compositions and relate drawings to a written method.

EUREKA
MATH™

Name _____ Date _____

1. Solve vertically. Draw chips on the place value chart and bundle, when needed.

   a. 65 + 75 = _____

100's	10's	1's

   b. 84 + 29 = _____

100's	10's	1's

   c. 91 + 19 = _____

100's	10's	1's

Lesson 21: Use math drawings to represent additions with up to two compositions and relate drawings to a written method.

87

©2015 Great Minds. eureka-math.org
G2-M4-SE-B2-1.3.1-1.2016

d. 163 + 27 = _____

100's	10's	1's

2. Abby solved 99 + 99 on her place value chart and in vertical form, but she got an incorrect answer. Check Abby's work, and correct it.

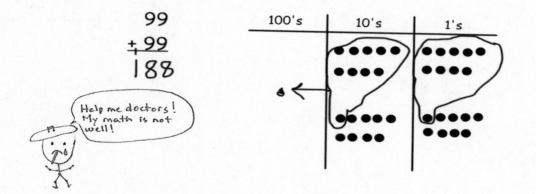

What did Abby do correctly?

_____

_____

What did Abby do incorrectly?

_____

_____

Lesson 21:   Use math drawings to represent additions with up to two compositions and relate drawings to a written method.

Name _____   Date _____

1. Solve vertically. Draw chips on the place value chart and bundle, when needed.

   a. 45 + 76 = _____

100's	10's	1's

   b. 62 + 89 = _____

100's	10's	1's

   c. 97 + 79 = _____

100's	10's	1's

EUREKA MATH

Lesson 21:   Use math drawings to represent additions with up to two compositions and relate drawings to a written method.

©2015 Great Minds. eureka-math.org
G2-M4-SE-B2-1.3.1-1.2016

89

d. 127 + 78 = _____

100's	10's	1's

2. The blue team scored 37 fewer points than the white team. The blue team scored 69 points.

   a. How many points did the white team score?

   b. How many points did the blue and white teams score altogether?

Lesson 21:   Use math drawings to represent additions with up to two compositions and relate drawings to a written method.

©2015 Great Minds. eureka-math.org
G2-M4-SE-B2-1.3.1-1.2016

Name _____   Date _____

1. Look to make 10 ones or 10 tens to solve the following problems using place value strategies.

a. 5 + 5 + 7= _____	25 + 25 + 17= _____	125 + 25 + 17= _____
b. 4 + 6 + 5 = _____	24 + 36 + 75 = _____	24 + 36 + 85 = _____
c. 2 + 4 + 8 + 6 = _____	32 + 24 + 18 + 46 = _____	72 + 54 + 18 + 26 = _____

Lesson 22:   Solve additions with up to four addends with totals within 200 with and without two compositions of larger units.

91

©2015 Great Minds. eureka-math.org
G2-M4-SE-B2-1.3.1-1.2016

2. Josh and Keith have the same problem for homework: 23 + 35 + 47 + 56. The
    students solved the problem differently but got the same answer.

Josh's work

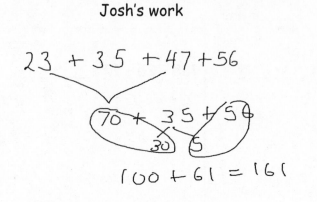

Keith's work

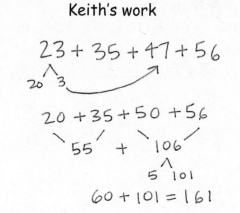

Solve 23 + 35 + 47 + 56 another way.

3. Melissa bought a dress for $29, a purse for $15, a book for $11, and a hat for $25.
    How much did Melissa spend? Show your work.

92   Lesson 22:   Solve additions with up to four addends with totals within 200 with
                            and without two compositions of larger units.

©2015 Great Minds. eureka-math.org
G2-M4-SE-B2-1.3.1-1.2016

Name _____   Date _____

1. Look to make 10 ones or 10 tens to solve the following problems using place value strategies.

a. 6 + 3 + 7= _____	36 + 23 + 17= _____	126 + 23 + 17= _____
b. 8 + 2 + 5 = _____	38 + 22 + 75 = _____	18 + 62 + 85 = _____
c. 9 + 4 + 1 + 6 = _____	29 + 34 + 41 + 16 = _____	81 + 34 + 19 + 56 = _____

EUREKA MATH™

Lesson 22:   Solve additions with up to four addends with totals within 200 with and without two compositions of larger units.

93

©2015 Great Minds. eureka-math.org
G2-M4-SE-B2-1.3.1-1.2016

2. The table shows the top six soccer teams and their total points scored this season.

Teams	Points
Red	29
Yellow	38
Green	41
Blue	76
Orange	52
Black	24

a. How many points did the yellow and orange teams score together?

b. How many points did the yellow, orange, and blue teams score together?

c. How many points did the red, green, and black teams score together?

d. Which two teams scored a total of 70 points?

e. Which two teams scored a total of 100 points?

Lesson 22:   Solve additions with up to four addends with totals within 200 with
                         and without two compositions of larger units.

©2015 Great Minds. eureka-math.org
G2-M4-SE-B2-1.3.1-1.2016

Name _____   Date _____

1. Solve using number bonds to subtract from 100. The first one has been done for you.

a. $106 - 90 = 16$ 6 100 $100 - 90 = 10$ $10 + 6 = \quad 16$	b. $116 - 90$
c. $114 - 80$	d. $115 - 80$
e. $123 - 70$	f. $127 - 60$

Lesson 23:   Use number bonds to break apart three-digit minuends and subtract from the hundred.

95

©2015 Great Minds. eureka-math.org
G2-M4-SE-B2-1.3.1-1.2016

g. 119 – 50	h. 129 – 60
i. 156 – 80	j. 142 – 70

2. Use a number bond to show how you would take 8 tens from 126.

Lesson 23:   Use number bonds to break apart three-digit minuends and subtract
                           from the hundred.

Name _____   Date _____

1. Solve using number bonds to subtract from 100. The first one has been done for you.

a.   105 – 90 = 15
 	
    /\\
 	
100   5
 	
    100 – 90 = 10
    10 + 5 = 15
	b. 121 – 90

c. 112 – 80	d. 135 – 70
e. 136 – 60	f. 129 – 50

EUREKA
MATH™

Lesson 23:   Use number bonds to break apart three-digit minuends and subtract from the hundred.

97

©2015 Great Minds. eureka-math.org
G2-M4-SE-B2-1.3.1-1.2016

g. 156 – 80	h. 138 – 40

2. Monica incorrectly solved 132 – 70 to get 102. Show her how to solve it correctly.

Monica's work:	Correct way to solve 132 – 70:
132 - 70 = _____ 100 32 100 - 30 = 70 70 + 32 = 102	

3. Billy sold 50 fewer magazines than Alex. Alex sold 128 magazines. How many magazines did Billy sell? Solve using a number bond.

Lesson 23:   Use number bonds to break apart three-digit minuends and subtract from the hundred.

©2015 Great Minds. eureka-math.org
G2-M4-SE-B2-1.3.1-1.2016

Name _____   Date _____

1. Solve using mental math. If you cannot solve mentally, use your place value chart and place value disks.

   a. 25 – 5 = _____ 25 – 6 = _____ 125 – 25 = _____ 125 – 26 = _____

   b. 160 – 50 = _____ 160 – 60 = _____ 160 – 70 = _____

2. Solve using your place value chart and place value disks. Unbundle the hundred or ten when necessary. Circle what you did to model each problem.

a.	b.
124 – 60 = _____	174 – 58 = _____
I unbundled the hundred. Yes No I unbundled a ten. Yes No	I unbundled the hundred. Yes No I unbundled a ten. Yes No
c.	d.
121 – 48 = _____	125 – 67 = _____
I unbundled the hundred. Yes No I unbundled a ten. Yes No	I unbundled the hundred. Yes No I unbundled a ten. Yes No
e.	f.
145 – 76 = _____	181 – 72 = _____
I unbundled the hundred. Yes No I unbundled a ten. Yes No	I unbundled the hundred. Yes No I unbundled a ten. Yes No

Lesson 24:   Use manipulatives to represent subtraction with decompositions of
1 hundred as 10 tens and 1 ten as 10 ones.

99

©2015 Great Minds. eureka-math.org
G2-M4-SE-B2-1.3.1-1.2016

g.

$111 - 99 = $ _____

I unbundled the hundred.   Yes   No
I unbundled a ten.   Yes   No

h.

$131 - 42 = $ _____

I unbundled the hundred.   Yes   No
I unbundled a ten.   Yes   No

i.

$123 - 65 = $ _____

I unbundled the hundred.   Yes   No
I unbundled a ten.   Yes   No

j.

$132 - 56 = $ _____

I unbundled the hundred.   Yes   No
I unbundled a ten.   Yes   No

k.

$145 - 37 = $ _____

I unbundled the hundred.   Yes   No
I unbundled a ten.   Yes   No

l.

$115 - 48 = $ _____

I unbundled the hundred.   Yes   No
I unbundled a ten.   Yes   No

3. There were 167 apples. The students ate 89 apples. How many apples were left?

Lesson 24:   Use manipulatives to represent subtraction with decompositions of
1 hundred as 10 tens and 1 ten as 10 ones.

©2015 Great Minds. eureka-math.org
G2-M4-SE-B2-1.3.1-1.2016

For early finishers:

4. Tim and John have 175 trading cards together. John has 88 cards.

    a. How many cards does Tim have?

    b. Brady has 29 fewer cards than Tim. Have many cards does Brady have?

Lesson 24:   Use manipulatives to represent subtraction with decompositions of
1 hundred as 10 tens and 1 ten as 10 ones.

©2015 Great Minds. eureka-math.org
G2-M4-SE-B2-1.3.1-1.2016

101

Name _____   Date _____

1. Solve using mental math. If you cannot solve mentally, use your place value chart and place value disks.

   a. 38 – 8 = _____   38 – 9 = _____   138 – 38 = _____   138 – 39 = _____

   b. 130 – 20 = _____   130 – 30 = _____   130 – 40 = _____

2. Solve using your place value chart and place value disks. Unbundle the hundred or ten when necessary. Circle what you did to model each problem.

a. 115 – 50 = _____	b. 125 – 57 = _____
I unbundled the hundred. Yes No I unbundled a ten. Yes No	I unbundled the hundred. Yes No I unbundled a ten. Yes No
c. 88 – 39 = _____	d. 186 – 39 = _____
I unbundled the hundred. Yes No I unbundled a ten. Yes No	I unbundled the hundred. Yes No I unbundled a ten. Yes No
e. 162 – 85 = _____	f. 172 – 76 = _____
I unbundled the hundred. Yes No I unbundled a ten. Yes No	I unbundled the hundred. Yes No I unbundled a ten. Yes No

g. 121 – 89 = _____ I unbundled the hundred. Yes No I unbundled a ten. Yes No	**h.** 131 – 98 = _____ I unbundled the hundred. Yes No I unbundled a ten. Yes No
i. 140 – 65 = _____ I unbundled the hundred. Yes No I unbundled a ten. Yes No	**j.** 150 – 56 = _____ I unbundled the hundred. Yes No I unbundled a ten. Yes No
k. 163 – 78 = _____ I unbundled the hundred. Yes No I unbundled a ten. Yes No	**l.** 136 – 87 = _____ I unbundled the hundred. Yes No I unbundled a ten. Yes No

3. 96 crayons in the basket are broken. The basket has 182 crayons. How many crayons are not broken?

Lesson 24:   Use manipulatives to represent subtraction with decompositions of
1 hundred as 10 tens and 1 ten as 10 ones.

103

©2015 Great Minds. eureka-math.org
G2-M4-SE-B2-1.3.1-1.2016

This page intentionally left blank

Name _____ Date _____

1. Solve the following problems using the vertical form, your place value chart, and place value disks. Unbundle a ten or hundred when necessary. Show your work for each problem.

a. 72 – 49	b. 83 – 49
c. 118 – 30	d. 118 – 85
e. 145 – 54	f. 167 – 78
g. 125 – 87	h. 115 – 86

2. Mrs. Tosh baked 160 cookies for the bake sale. She sold 78 of them. How many cookies does she have left?

3. Tammy had $154. She bought a watch for $86. Does she have enough money left over to buy a $67 bracelet?

Lesson 25:   Relate manipulative representations to a written method.

©2015 Great Minds. eureka-math.org
G2-M4-SE-B2-1.3.1-1.2016

EUREKA
MATH™

Name _____   Date _____

1. Solve the following problems using the vertical form, your place value chart, and place value disks. Unbundle a ten or hundred when necessary. Show your work for each problem.

a. 65 – 38	b. 66 – 49
c. 111 – 60	d. 120 – 67
e. 163 – 66	f. 184 – 95
g. 114 – 98	h. 154 – 85

2. Dominic has $167. He has $88 more than Mario. How much money does Mario have?

3. Which problem will have the same answer as 133 – 77? Show your work.

   a. 155 – 66

   b. 144 – 88

   c. 177 – 33

   d. 139 – 97

Lesson 25: Relate manipulative representations to a written method.

EUREKA MATH

©2015 Great Minds. eureka-math.org
G2-M4-SE-B2-1.3.1-1.2016

Name _____   Date _____

1. Solve vertically. Draw chips on the place value chart. Unbundle when needed.

   a.   181 – 63 = _____

hundreds	tens	ones

   b.   134 – 52 = _____

hundreds	tens	ones

   c.   175 – 79 = _____

hundreds	tens	ones

EUREKA
MATH™

Lesson 26:   Use math drawings to represent subtraction with up to two
              decompositions and relate drawings to a written method.   109

©2015 Great Minds. eureka-math.org
G2-M4-SE-B2-1.3.1-1.2016

d.   115 – 26 = _____

hundreds	tens	ones

e.   110 – 74 = _____

hundreds	tens	ones

2. Tanisha and James drew models on their place value charts to solve this problem:
    102 – 47. Tell whose model is incorrect and why.

James   Tanisha

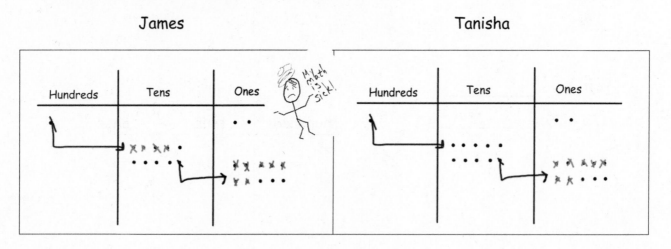

_____'s model is incorrect because _____

_____.

Use math drawings to represent subtraction with up to two
decompositions and relate drawings to a written method.

Name _____   Date _____

1. Solve vertically. Draw chips on the place value chart. Unbundle when needed.

    a.   114 – 65 = _____

hundreds	tens	ones

    b.   120 – 37 = _____

hundreds	tens	ones

    c.   141 – 89 = _____

hundreds	tens	ones

  EUREKA MATH   **Lesson 26:**   Use math drawings to represent subtraction with up to two   **111**
decompositions and relate drawings to a written method.

©2015 Great Minds. eureka-math.org
G2-M4-SE-B2-1.3.1-1.2016

d. 136 – 77 = _____

hundreds	tens	ones

e. 154 – 96 = _____

hundreds	tens	ones

2. **Extension**: Fill in the missing number to complete the problem. Draw a place value chart and chips to solve.

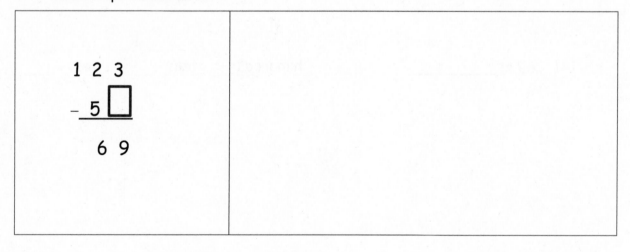

```
  1 2 3
 -  5 □
   ─────
    6 9
```

Lesson 26:   Use math drawings to represent subtraction with up to two
decompositions and relate drawings to a written method.

©2015 Great Minds. eureka-math.org
G2-M4-SE-B2-1.3.1-1.2016

Name _____   Date _____

1. Make each equation true.

    a. 1 hundred = _____ tens

    b. 1 hundred = 9 tens _____ ones

    c. 2 hundreds = 1 hundred _____ tens

    d. 2 hundreds = 1 hundred 9 tens _____ ones

2. Solve vertically. Draw chips on the place value chart. Unbundle when needed.

    a. 100 – 61 = _____

hundreds	tens	ones

    b. 100 – 79 = _____

hundreds	tens	ones

EUREKA MATH™

Lesson 27: Subtract from 200 and from numbers with zeros in the tens place.

113

c. 200 – 7 = _____

hundreds	tens	ones

d. 200 – 87 = _____

hundreds	tens	ones

e. 200 – 126 = _____

hundreds	tens	ones

Lesson 27:   Subtract from 200 and from numbers with zeros in the tens place.

©2015 Great Minds. eureka-math.org
G2-M4-SE-B2-1.3.1-1.2016

Name _____   Date _____

1. Solve vertically. Draw chips on the place value chart. Unbundle when needed.

   a.   100 – 37 = _____

hundreds	tens	ones

   b.   100 – 49 = _____

hundreds	tens	ones

   c.   200 – 49 = _____

hundreds	tens	ones

EUREKA MATH™

Lesson 27:   Subtract from 200 and from numbers with zeros in the tens place.

115

©2015 Great Minds. eureka-math.org
G2-M4-SE-B2-1.3.1-1.2016

d. 200 – 57 = _____

hundreds	tens	ones

e. 200 – 83 = _____

hundreds	tens	ones

2. Susan solved 200 – 91 and decided to add her answer to 91 to check her work.
    Explain why this strategy works.

Susan's work:	Explanation:
$$\begin{array}{r} {}^{1}\cancel{2}\ {}^{9}\cancel{0}\ {}^{10}\cancel{0} \\ -\ \ 9\ 1 \\ \hline 1\ 0\ 9 \end{array} \qquad \begin{array}{r} 1\ 0\ 9 \\ +\ \ 9\ 1 \\ \hline 2\ 0\ 0 \end{array}$$	_____ _____ _____ _____

Lesson 27:   Subtract from 200 and from numbers with zeros in the tens place.

Name _____ Date _____

1. Solve vertically. Draw chips on the place value chart. Unbundle when needed.

    a. 109 – 56 = _____

hundreds	tens	ones

    b. 103 – 34 = _____

hundreds	tens	ones

    c. 200 – 155 = _____

hundreds	tens	ones

EUREKA
MATH™

Lesson 28:   Subtract from 200 and from numbers with zeros in the tens place.

117

©2015 Great Minds. eureka-math.org
G2-M4-SE-B2-1.3.1-1.2016

d. 200 – 123 = _____

hundreds	tens	ones

2. Solve vertically without a place value chart.

200 – 148 = _____

3. Solve vertically. Draw a place value chart and chips.

Ralph has 137 fewer stamps than his older brother. His older brother has 200 stamps. How many stamps does Ralph have?

Lesson 28:   Subtract from 200 and from numbers with zeros in the tens place.

Name _____ Date _____

1. Solve vertically. Draw chips on the place value chart. Unbundle when needed.

   a.   136 – 94 = _____

hundreds	tens	ones

   b.   105 – 57 = _____

hundreds	tens	ones

   c.   200 – 61 = _____

hundreds	tens	ones

EUREKA
MATH™

Lesson 28:   Subtract from 200 and from numbers with zeros in the tens place.

119

d. 200 – 107 = _____

hundreds	tens	ones

e. 200 – 143 = _____

hundreds	tens	ones

2. Herman collected 200 shells on the beach. Of those, he kept 136 shells and left the rest on the beach. How many shells did he leave on the beach?

Lesson 28: Subtract from 200 and from numbers with zeros in the tens place.

Name _____ Date _____

1. Solve each addition expression using both the totals below and new groups below methods. Draw a place value chart with chips and two different number bonds to represent each.

a. 27 + 19

New Groups Below	Totals Below	Place Value Chart	Number Bonds

b. 57 + 36

New Groups Below	Totals Below	Place Value Chart	Number Bonds

Lesson 29: Use and explain the totals below method using words, math drawings, and numbers.

121

©2015 Great Minds. eureka-math.org
G2-M4-SE-B2-1.3.1-1.2016

2. Add like units and record the totals below.

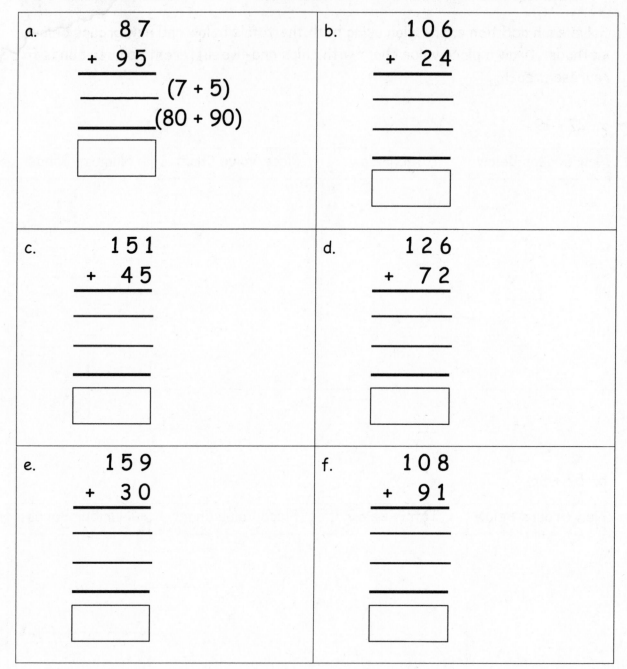

a.
```
    87
+   95
```
_____ (7 + 5)

_____ (80 + 90)

b.
```
   106
+   24
```

c.
```
   151
+   45
```

d.
```
   126
+   72
```

e.
```
   159
+   30
```

f.
```
   108
+   91
```

Lesson 29:   Use and explain the totals below method using words, math drawings, and numbers.

©2015 Great Minds. eureka-math.org
G2-M4-SE-B2-1.3.1-1.2016

Name _____   Date _____

1. Add like units and record the totals below.

a.	b.
48 + 27 ——— ——— [ ]	118 + 73 ——— ——— ——— [ ]

c.	d.
156 + 62 ——— ——— ——— [ ]	137 + 82 ——— ——— ——— [ ]

EUREKA
MATH™

Lesson 29:   Use and explain the totals below method using words, math drawings, and numbers.

123

©2015 Great Minds. eureka-math.org
G2-M4-SE-B2-1.3.1-1.2016

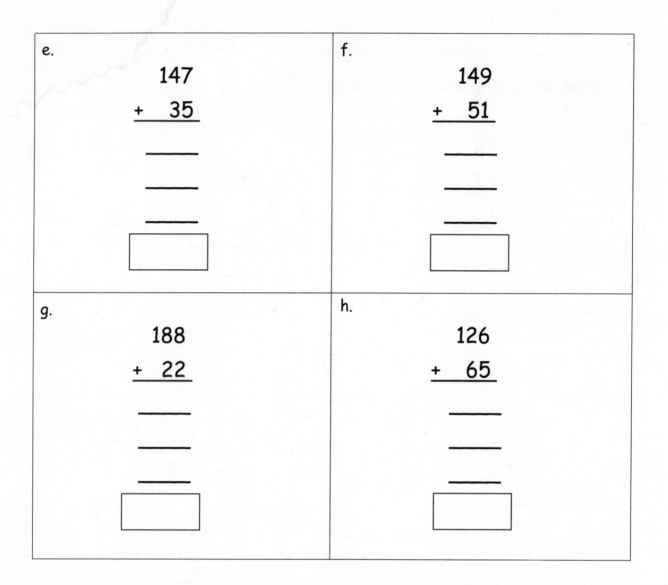

e.

$$147$$
$$+ \ \ 35$$

_____

_____

_____

f.

$$149$$
$$+ \ \ 51$$

_____

_____

_____

g.

$$188$$
$$+ \ \ 22$$

_____

_____

_____

h.

$$126$$
$$+ \ \ 65$$

_____

_____

_____

2. Daniel counted 67 apples on one tree and 79 apples on another tree. How many apples were on both trees? Add like units and record the totals below to solve.

Name _____ Date _____

1. Linda and Keith added 127 + 59 differently. Explain why Linda's work and Keith's work are both correct.

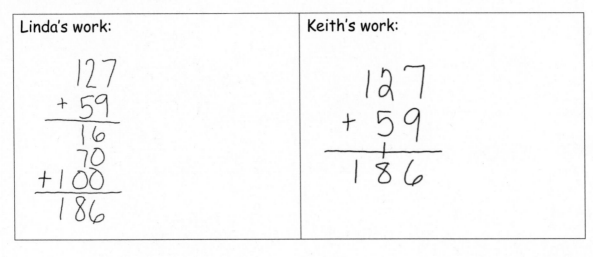

Linda's work:	Keith's work:
127 + 59 ――― 16 70 +100 ――― 186	127 + 59 ――― 186

_____

_____

_____

2. Jake solved 124 + 69 using new groups below. Solve the same problem another way.

124
+ 69
―――
193

Lesson 30: Compare totals below to new groups below as written methods.

125

©2015 Great Minds. eureka-math.org
G2-M4-SE-B2-1.3.1-1.2016

3. Solve each problem two different ways.

a. 134 + 48	b. 83 + 69
c. 46 + 75	d. 63 + 128

Name _____   Date _____

1. Kari and Marty solved 136 + 56.

Kari's work:	Marty's work:
$$\begin{array}{r} 136 \\ +\ 56 \\ \hline 192 \end{array}$$	$$\begin{array}{r} 136 \\ +\ 56 \\ \hline 12 \\ 80 \\ +100 \\ \hline 192 \end{array}$$

Explain what is different about how Kari and Marty solved the problem.

_____

_____

_____

2. Here is one way to solve 145 + 67. For (a), solve 145 + 67 another way.

	a.
$\begin{array}{r} 1\,4\,5 \\ +\ \ 6\,7 \\ \scriptstyle 1\ 1 \\ \hline 2\,1\,2 \end{array}$	

b. Explain how the two ways to solve 145 + 67 are similar.

_____

_____

_____

3. Show another way to solve 142 + 39.

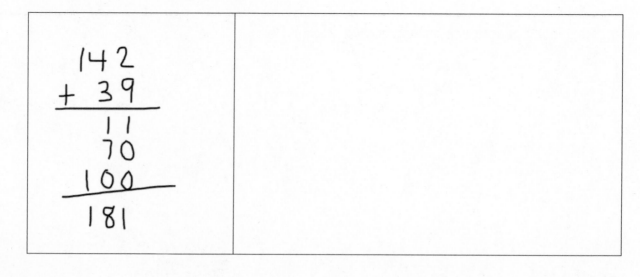

$\begin{array}{r} 1\,4\,2 \\ +\ 3\,9 \\ \hline 1\,1 \\ 7\,0 \\ 1\,0\,0 \\ \hline 1\,8\,1 \end{array}$

Lesson 30:   Compare totals below to new groups below as written methods.

Name _____   Date _____

Solve the following word problems by drawing a tape diagram. Use any strategy you have learned to solve.

1. Mr. Roberts graded 57 tests on Friday and 43 tests on Saturday. How many tests did Mr. Roberts grade?

2. There are 54 women and 17 fewer men than women on a boat.

    a. How many men are on the boat?

    b. How many people are on the boat?

3. Mark collected 27 fewer coins than Craig. Mark collected 58 coins.

    a. How many coins did Craig collect?

    b. Mark collected 18 more coins than Shawn. How many coins did Shawn collect?

4. There were 35 apples on the table. 17 of the apples were rotten and were thrown out. 9 apples were eaten. How many apples are still on the table?

Lesson 31:   Solve two-step word problems within 100.

Name _____ Date _____

1. Melissa had 56 pens and 37 more pencils than pens.

   a. How many pencils did Melissa have?

   b. How many pens and pencils did Melissa have?

2. Antonio gave 27 tomatoes to his neighbor and 15 to his brother. He had 72 tomatoes before giving some away. How many tomatoes does Antonio have left?

3. The bakery made 92 muffins. Seventeen were blueberry, 23 were cranberry, and the rest were chocolate chip. How many chocolate chip muffins did the bakery make?

4. After spending $43 on groceries and $19 on a book, Mrs. Groom had $16 left. How much money did Mrs. Groom have to begin with?